Traditional British Honey Drinks

Francis Beswick

www.northernbeebooks.co.uk
Published by Northern Bee Books 2015
Scout Bottom Farm
Mytholmroyd
Hebden Bridge HX7 5JS (UK)

Traditional British Honey Drinks
Francis Beswick

©Text copyright F.C.Beswick 1992, 1994,2003
© Illustrations copyright David Taylor 1994, 2003

First published as
Traditional Honey Drinks
by author 1992

All rights reserved. No part of this publication my be reproduced, stored in a retrieval system, transmitted in any form or by any means electronic, mechanical, including photocopying, recording or otherwise without prior consent of the copyright holder

ISBN 978-1-908904-78-2

Layout: gustav moody
Printed by Lightning Source, UK

www.northernbeebooks.co.uk

Published by Northern Bee Books 2015
Scout Bottom Farm
Mytholmroyd
Hebden Bridge HX7 5JS (UK)

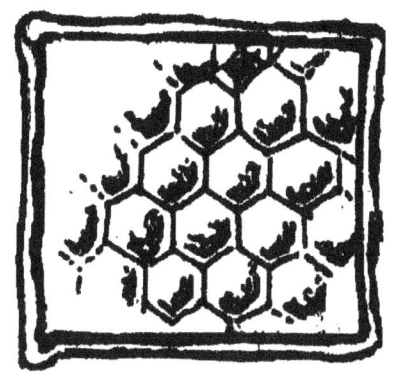

There is in British tradition a surprising and almost totally neglected variety of alcoholic drinks. Invariably, many, such as cock ale, a drink made from immersing dates, raisins and a dead cock in an ale vat, will have fallen out of favour for good reasons, but along with them there has nearly vanished a number of very attractive brews which are now made only by handfuls of enthusiasts. Honey brews seem to have suffered more than most, and mead, one of the oldest drinks available to mankind, practically disappeared in much of the Western World at one time, although it is now making a limited recovery.

Industrialization was the cause of this narrowing. Mead had been brewed in many country homes upto the eighteenth century, but in the late eighteenth there came population growth and a drift to the cities. To feed a growing population much greater acreage of land went under intense cereal production, and the economics of size meant that barley for beer became cheaper than honey for mead. Even then, mead was still a popular drink, and was part of the rations of British troops fighting Napoleon in Egypt. The nail in the coffin of mead came with the availability of cane and beet sugar, the former from the West Indies and the latter from the sugar beet, from which scientists had finally been able to extract the sucrose. Sugar quickly outstripped honey as a sweetener, and honey production declined to the point at which honey became something of a luxury rather than a staple. Mead practically disappeared with it and became the preserve of a few remaining traditionalists.

Yet honey brews had enjoyed a long and distinguished history. Probably the first mead was made when some prehistoric folk mixed honey with water and saw it ferment, the result of the wild yeasts which were found on the surface of the honey. Honey ferments came to be credited with special powers, especially for fertility. The Saxons and the Norse used to drink mead at wedding feasts, which lasted a month, known as a moon, giving us the word honeymoon, the month in which honey drinks were taken. The mead was thought to have the power of enhancing the fertility of the couple and it was sometimes the case that after the bride went to bed the groom was filled with mead to boost his virility for the night's endeavours. That he was still capable after drinking so much alcohol might suggest that mead has something to commend it!

One of the special powers attributed by many peoples at various times in history to honey brews was that of maintaining health and vigour. This has a clear connection with fertility, for ultimately a man's virility is linked to his overall physical condition. That this belief is widespread indicates that there is some truth in it. However, we must not look for elixirs of youth or magical ingredients. The truth is more mundane. Wild honeys are often the product of the pollen of many flowers, and as such contain a wide variety of trace elements, which are essential to good health. At a time when tonics were unavailable and scientifically planned diets were non-existent, a honey drink was the nearest to a tonic or a diet supplement mankind could reach.

The Celts were as keen on honey brews as were the Saxons. Taliesin sings the virtues of mead, a distilled, and, therefore, strong variety of the drink. A Dr Howells of Oxford writing in the eighteenth century asserted that the Celtic druids and bards used to take the drink before engaging in speculations, but he provides us with no clues as to where he discovered this information. That one form of mead, metheglynn, is known by a Welsh name indicates that mead was highly valued by the Celtic peoples of these islands.

Mead, especially its pure form made with honey alone, was very much a festal drink. This was to a great degree due to its price. A good mead is not made quickly and requires honey, which has never been cheap. The honeymoon has already been mentioned, but there were other feasts. The great midwinter festival, which we celebrate as Christmas, but whose roots go deep into the pre-Christian past, would have been one such festival. This was always celebrated with great carousing in Northern and Western Europe. It still is. Midwinter is a time when we celebrate fertility and rebirth. Winter, the season of death, reaches its deepest and darkest point, and then the cycle swings towards the warmth of the coming spring. Many of our symbols of Christmas refer to the renewal of life: the evergreen Christmas tree, which never sheds its leaves, the holly, which bears fruit in the coldest season, the mistletoe, which draws the strength of the oak in all seasons of the year. For a midwinter feast mead, the drink of fertility would have been particularly apt. It would have been the drink which would have been saved for the special occasion. This is not to say that other drinks were not consumed, far from it, but mead would have been the drink for the occasion.

The origin of the significance given to honey and mead may go back to the neolithic period, in which the mother goddess was depicted as a bee. In Minoan Crete, about 4000 BC, the goddess and her priestesses are shown on a seal stone dressed as bees and dancing together. The hive was presumably her womb. Thus, honey, the fruit of the womb, was a divine gift to man and was intimately connected with worship. Thus, its brewing into ritual drinks was a natural development. It may have served as a kind of sacrament of the goddess.

There are tales which indicate that mead was credited with magical powers. In the Icelandic Prose Edda we are told of the origin of the mead of poetry. The story is that the wise and widely travelled Kvasir, who wandered the world dispersing wisdom, was treacherously murdered by two dwarfs, who mingled his blood with honey and made mead. Whoever drank from the blood and honey cup became a poet and a scholar. Such legends invariably mediate to us a tradition from a long-gone and misty past. Was there a time when people used a honey-blood mixture in religious rites, perhaps to gain inspiration. Was this a memory of brews taken by shamans in palaeolithic times? Who knows. Certainly in the myth the brew is taken by Odin, who became the patron of poets. Such accounts of a deity's victory over dwarfs or titans etc usually conceal the conquering or domination of one culture by another, and certain of the Norse deities, e.g. Frey, are thought to be the deities of an earlier culture. This distant culture may have blended blood and honey in some religious rites, possibly to provide inspiration for its seers or shamans. The evidence suggests it.

The Process of Mead Making

The first step in mead making was to collect the honey. Tradition has it that the best meads came from the honey first drained from the hive, for at that time it is not only still liquid but is unlikely to have much wax in it. There was a variety of kinds of honey. Some, such as heather and clover honey, were highly prized for flavour. Rosemary and wild rose honey were nearly as prized, but were probably less widely used, as these plants are less common than heather and clover. There was also a wide range of mixed honeys, taken from bees which fed on a variety of plants. Such polyflora honeys had flavours less distinctive than did the monoflora ones.

The honey was mixed with rain or spring water and boiled gently. The Romans were said by Pliny to have used rain water, but spring water was pure and mineral rich, and would have added character to the mead. The boiling was to eliminate impurities, although too vigorous a boiling would destroy some of the volatile flavourings. Impurities would be skimmed off. What was left was the must from which the mead was brewed. The rule of thumb for the strength of this must was that you had to be able to float an egg on it, and that the egg should not sink to more or less than half its depth.

Some of the must was drained off into a bottle, into which was added a brewer's yeast. This became the fermentation starter to be added to the must when it was sufficiently active. Fermentation took place in stone vessels which were kept in places with a warm, stable temperature. When fermentation was finished the must was transferred to stone vessels and stored in a cool place until mature. Mead is not a quick drink. A year was considered normal, and heather mead was sometimes matured for seven years, a custom that has given rise to the misconception that this length of maturation is necessary for all meads. A year's maturation is sufficient.

A variety of drinks could be produced. The mead could be of either ale or wine strength. Wine mead is the only form of mead commonly drunk today, but the Saxons had ale mead, which they knew as *alu*, a drink of ale strength. Mead could also be sweet or dry. It will be sweet if some of the honey sugars remain, but if the fermentation consumes all the sugar then the resulting drink will be dry. That mead was described by some enthusiasts as 'fierce, hot and dry' indicates that it is possible to consume all the sugars, if the must is of the right strength and the yeast is of a strong kind. This is more likely to happen with a wine yeast than a brewer's yeast, for the latter ceases fermentation earlier than the former. Distilled meads were known among the Celts, as I said earlier, but beyond what is known of early distillation techniques, nothing is known of the details of the process or of the character of the drink produced.

That honey was considered to be healthgiving led to its having a medicinal use. A native British product with a medicinal function was **metheglynn**, which is a term derived from the Welsh word for medicine. The reasoning behind this use was that herbs were healthy, and so was honey, therefore a combination of herbs and honey would have a beneficial effect greater than either would have singly. Certainly, metheglynn was regarded in the mediaeval age as a panacaea for many ills. There is no hard and fast rule as to which herbs were used. The Celts often infused mead with a juice extracted from the bark of the hazel tree, which was sacred to them. Fennel would be another favourite, as it was a herb with wide medicinal usage. But often mixed herbs would be used. It would, of course, have been possible for a healer to specify a metheglynn made with a specific herb, and it may have been done, but metheglynn takes so long to make that there would be no point in prescribing it for a patient. Mixed herbs seem to have been the most economical way of using both herbs and honey.

During the Mediaeval period a strong metheglynn known as **sack metheglynn** was produced in Wales. It was not only strong but sweet. As strong meads have usually used up the sugar content of the must it seems likely that a strong and sweet drink could only be produced by distillation, which was, as we have seen, known to the Celts. Alchemical thought, which was in vogue in the mediaeval period, may have had some part to play in the development of sack metheglynn. Alchemists has distilled wine to reach its essence, that which made it what it was. The thinking was that if the dilute substance was beneficial, the undilute form would be more beneficial still. Mixed with herbs it would be powerful indeed.

Another medicinal kind popular in Britain was **hippocras**. This was said to have been invented by Hippocrates, the ancient Greek physician. It was a brew of honey and white wine to which herbs had been added.

Later forms of metheglynn and hippocras began to use imported spices, either in addition to or in place of herbs. This would have occurred after the Dark Ages as trade with the East began to develop and such exotic delicacies as cloves, cinnamon and nutmeg began to find their way onto the tables of the rich. Spices would not seem to be medicinal to the modern mind, but to the mediaevals they were simply herbs from an exotic location and would be credited with herbal powers.

Metheglynn and hippocras take longer to mature than mead does, for herbs and spices do not easily mix with the mead, and so blending requires patience. It was, of course, possible to use water in which herbs had already been infused to brew the mead, and this would have taken less time than would have been taken had the herbs been allowed to dissolve. Another quicker method would have been to have suspended the herbs in a bag for some time in the must. However, metheglynn is notoriously hard to clear. In early times egg shells, crushed and dropped into the fermented must, would have been used, whereas nowadays finings would be available.

Mixed Honey Drinks

Honey has never been cheap, and so it was considered economically prudent to spread it out. A range of brews made from honey and another substance came into use. Some of these are **melomels**, which are fermented from honey and fruit juice. Two have a long history. These are **red mead** and **black mead**, which are fermented from a brew of red and black currants respectively, although strictly speaking the name mead should not be given to them, for it should only apply to a drink made purely from honey. Any light fruit will be useful for melomel making. Traditionally, gooseberries would have been available to British brewers. It is possible that pear melomel could have been made, although I have found no recipes for it. Raspberries also have been used to brew an attractive melomel. Traditional country melomel brewers avoided elderberries and blackberries, as their tannin content is too high and makes for an astringent taste to the finished product.

A practically-lost kind of ale-strength melomel was **cyser**. This was a mixture of apple juice and honey, and it was very commonly made in monasteries, which kept apple orchards and beehives, the latter of which were necessary to maintain the supply of beeswax, which the Roman Catholic Church enacts to be the material used for making candles for use on the altar. Monasteries often had extensive orchards. Surplus apples and honey could be put to use in the making of cyser, which was drunk in the less severe monasteries by the monks and in the more severe ones in the guest house. In theory an equivalent of cyser made with pears is possible, but there is no evidence that it was ever produced.

Others combined honey and malt in equal quantities into a brew alternatively known as **braggot or braggon**. Braggon is mentioned favourably by Chaucer, and was very popular in the thirteenth century. It continued to be brewed in rural districts upto the late eighteenth, when it faded away. It is likely that a substantial number of types of this brew existed, for the English were wont to steep a variety of fruits in ale to add richness and flavour to the brew, and as braggot was a close relation of ale it is conceivable, indeed probable, that fruits were routinely added to a braggot mixture. There are, unfortunately, no surviving recipes which specify that this should be done.

This brings us to **wassail**. This was a Saxon greeting or toast roughly meaning 'good health'. The wassail bowl was drunk on Twelfth Night and Christmas Eve. There were a variety of wassail bowls. It seems that a concoction of ale, herbs, fruit and honey in the proportions that the brewer favoured were blended together. Spices have been included in wassail, but as spices were not easily available to the Saxons, whereas herbs were, it is probable that they were a later introduction.

Popular with the clergy was a class of imported drinks known as **piments**, and clerics were often criticized for their excessive taste for them, so popular that in 817 a local synod at Aix la Chapelle tried to ban the clergy from drinking spiced wines. The prohibition was observed less than the law on celibacy was, and as clerical marriage was common, the law was probably honoured in the breach rather than the observance. Piments were a concoction of white wine and honey fermented together and flavoured with spices and herbs. In England piments were most probably imports, although for a time south-west England had a minor wine industry which survived until the conquest of south-west France by Edward the Third opened the ports to much high-quality French wine. It is possible that English grapes went not only to produce wine but also to produce local piments. This would have been a suitable use for what was essentially a lower quality grape crop which would not have provided anything better than a mediocre table wine. No records survive to support this claim, although it remains a well-grounded possibility.

In more modern times there developed a range of drinks known as **honey wines and beers**. The difference between these and wine and ale meads is not clear, but they are usually concoctions of grape juice and honey fermented for a short time to make a drink for Summer afternoons. The ferment is effected by heating the must, rather as is done in distilling, and a quick ferment is effected, but the honey, which is slower in fermenting than is the grape juice, probably contributes little to the alcoholic strength of the drink. Honey beer is a drink of ale strength produced in this way, but its distinct characteristic is that it is, like ginger beer, meant to be drunk while fermenting. In such brews honey is to be regarded mainly as a sweetener, and they are not the sort of drinks which would ever be stored long or acquire a vintage.

Not all drinks used fermented honey. One class of winter warmers known as **bishops** used honey as a sweetener. These probably arose in the mediaeval period, but became commoner in post-mediaeval times as citrus fruits and spices became more easily available through easern trade In the modern centrally-heated age when transport is swift we rarely have to face the real problems of winter travel in cold conditions. It is only when we are stranded in a blizzard that we come to see how unpleasant winter is. In earlier times people were closer to the dangers of winter. Thus, a coach traveller journeying through bad weather would arrive at an inn feeling in desperate need of warmth. In such conditions 'mine host' would have been able to offer a bishop. This was a brew formed from hot water, mulled wine, honey, an orange or lemon and various spices, probably whatever were available. Essentially the hot water, wine, honey and spices were mixed in a punch bowl into which was dropped the spiced citrus fruit. There was a variety of bishops which used different wines, fruits and spices. Some were made with red wines. However, if claret was used the bishop was named **cardinal**, to indicate its higher status. Even higher in status was **pope**, which could be made with a champagne or the rare Tokay. The rule for bishops seems to have been that a variety of kinds of wine, honey, spices and citrus fruits could be used in any combination. In the eighteenth century tavern regulars could have their own bishop made by the landlord. Recipes are really just advisory. They are what someone has enjoyed and are in no way exclusive of individual experiment and creativity.

A class of drinks closely related to the bishops was the **neguses**. These derive from Colonel Francis Negus, the Member of Parliament for Ipswich. It seems that Negus was somewhat disturbed by conditions in the House of Commons, where the temperature was quite low during a winter debate. Negus concocted a brew made from boiling wine, honey, spices and citrus rinds together with some water. This was intended to keep him warm while in the chamber. Really, neguses differ from bishops insofar as they use only the rind rather than the whole fruit and insofar as they boil the ingredients together rather than in separate pans, as bishops do. The rule for neguses is the same as for bishops. Recipes are advisory, and the kinds of wine, spices and fruit used are at the brewer's discretion.

One drink popular in the Elizabethan period was **caudle**. It was often used as a nightcap, and people were often 'caudled' to bed with it when cold or sick. It was also recommended for women in childbirth. Brown ale was poured over four ounces of honey and a little oatmeal, and the result was heated in a slow oven for some time, whereupon the beer was drained off and spiced with nutmeg and lemon juice (orange would be just as good). Later, whiskey or rum were added if desired. It is said to be an effective cure for insomnia. It was also possible to produce a caudle by boiling honey, water and spice together,then mixing them with an ale boiled separately. Again, the rule is as for bishops: choose your own ale and your own spices. The oatmeal seems to be common to such recipes.

The rule for brewing such drinks as bishops, neguses and caudles is that there are no rules. There are merely guidleines. Basically, a variety of kinds of honey, wine, spices and fruits may be mixed as the maker pleases. It is an area in which there is scope for originality and individual taste. Recipes are guidelines. They identify what has in the past been pleasing, but trial and error can be as effective in producing good results. Wine may be replaced by spirits. The Scots certainly invented a drink called **athol brose**, a concoction of whiskey and honey, which was considered by highlanders to be the ideal breakfast drink. It is possible that in Celtic times distilled mead was used in such winter warmers, as it would serve this purpose as whiskey now serves it, although there is no evidence of this.

Techniques

If you can make wine, you can make mead or any honey-based drink. The equipment is exactly the same, and the technique is not substantially different. Simply, instead of a must of grape or fruit juice you use a must of honey and whatever else you want to include. The equipment is what you would find in an amateur wine maker's kit, namely: two demijons (large jars), one for fermenting and the other for racking (letting the wine or mead mature after the fermentation has stopped); a syphon tube for extracting the liquid from its jar; a fermentation lock, pierced cork to insert the lock, yeast, a hydrometer and sufficient bottles and corks. Then you are ready to begin.

There are, however, some points to consider.

1: Not all honeys are suitable for brewing with, and some give less flavour than others do. The general principle is that the best flavour comes from an impure, monoflora honey. A monoflora honey is one produced by bees which have fed on only one kind of pollen, such as heather or clover. This is most likely to be possible in areas where one crop predominates. I feel that small-scale beekeepers are the ones most likely to provide such honey. Why impure? A commercially purified honey loses many of its trace elements and flavourings, which are an important part of its nutrient content and its taste. That a honey is impure in this way is not dangerous to health. Any bacterial impurities which might constitute a danger can be eliminated by the mead maker with no loss of useful flavourings, as I will explain below.

Some meadmakers have in the past found difficulty with some American, Canadian and Australian honeys. Eucalyptus honey is said to be useless for mead-making. As this book is about traditional honey-brews I am not concerned with those honeys, for my interest is with traditional brews alone.

This is not to decry polyflora honeys, for a rich variety of flavourings may be present in them, and you can make a pleasant mead, but purified honeys are mainly sugar. I think that the best mead comes from honeys produced by small-scale, traditional beekeepers. It is a good idea to get it straight from the hive, if possible, for at this stage no flavourings have had time to decay and no-one can have purified anything out.

The honey was traditionally taken from the hive by piercing the bottom to let the honey, liquid in its initial state, flow out, and the first honey from the hive was considered to be the most suitable for mead making. Honey solidifies as it ages, and modern liquid honeys have a chemical added to maintain their liquidity. Liquid honey, of course, dissolves more easily than does solid honey, and there has been less time for its fragrances to decay through age. Solid honey must be melted to liquidity, taking more heat and possibly losing some of its fragrance.

2: The must should be purified of bacteria. This was traditionally done by the light boiling traditionally practised by meadmakers. Although nothing was known of bacteria in mediaeval times experience showed that a lightly boiled mead tasted best. Nowadays boiling is generally unnecessary, as campden tablets dropped into the must prior to the addition of the yeast more than adequately kill the bacteria

Some meadmakers still practise light boiling when making heather mead or any brew using heather honey, for this eliminates some bitter elements in the brew. The boiling should be light and the must should not be brought to boiling point. Ten minutes of light boiling should suffice. Boiling would also be necessary should the honey contain wax. Wax is actually quite nourishing and digestible, but the meadmaker can do wthout it. Personally, I would rather not use a honey that contained any wax. It is too much trouble.

3: Traditionally the must should be of sufficient thickness to float an egg to no more than half its depth. This is important, for a must which is much thinner will produce a mead which has less body and is shorter in flavour.

4: The yeast must be chosen properly. There are no specific mead yeasts, but the best results seem to be achieved with white wine yeasts. The following yeasts are known to have been used with good results:

All purpose
Bernkastler
Bordeaux
Champagne
Graves
Liebfraumilch
Maury
Steinberg
Tokay
Zeltinger

Brewer's yeast is useful for ale meads and other drinks of ale-strength. Baker's yeast works but gives a poor flavour.

5: The acid content of the must is vitally important. Pure meads require between one-half to three-quarters of an ounce of acid per gallon. The favoured acid is often citric, for it produces a pleasant flavour. This was important in traditional mead making, for citric acid was available in citrus fruits, and the addition of a drop of lemon or orange was the means by which mead makers could add the necessary acid to their must. What they did in the British Isles before the importing of citrus is unknown. Presumably acidic herbs were added to the must in quantities discovered by trial and error and passed on by word of mouth from one generation to the next. For high quality meads mead-acid mixture is used. This is composed of two-thirds malic and one-third tannic acid. It is applied to the mixture in the same proportions as is citric acid.

For drinks made from honey with another substance add acid in the proportions which you would do for a wine.

Tannic acid is an essential element in a mead, for it provides astringency. It is added in the proportion of one-fifteenth to one-twentieth of an ounce per gallon to a pure mead must, although in melomel and other mixed musts the tannin requirement is so varied that it is wise to look at the individual recipe. Before tannic acid was purchased mead makers used to add astringent herbs, such as hops, and later it came to be known that cold tea in the proportion of a table spoon to a gallon of must provided the necessary astringency. Both elderberries and blackberries are rich in tannin, so it is more than likely that they were used in small quantities to add the necessary tannin to the must. Nowadays, tannic acid is, of course, purchasable anywhere wine-making equipment is sold.

6: The must requires the addition of a number of nutrients if it is to successfully ferment. Chief among these is ammonium phospate, and it is thought necessary that one level tablespoon of this substance be applied to every gallon of must, although as ammonium phospate comes in tablet

form the rule is that one level spoonful equals two tablets. Magnesium phospate (Epsom salts) is also useful and a pinch should be applied to every gallon of must.

7: Racking, transferring the brew from the fermentation jar to a storage jar, should take place within a week of fermentation finishing. A second racking should be performed when a thick sediment has settled. Further rackings every three or four months should be done if necessary, that is if further sedimentary deposits are found in the demijon. It is worth noting that with melomels it is possible to get a thick scum on the top, a phenomenon I have found with pineapple melomel. This should be eliminated by inserting the siphon tube below the level of the scum when racking. On each racking add a campden tablet to the liquid to protect against bacteria.

Mead making is a patient process. There are three week wines but no three week meads. Seven years is too long for all but the best heather meads, but I cannot think that longer than a year is necessary for wine meads. This is not to deny that a longer maturation can be beneficial, for some meadmakers intent on top quality meads mature for three years for a light mead and up to four for a fuller kind, but it is not an absolute necessity. Ale mead may take a shorter time, about as long as beer or ale. The maturation period for the other mead drinks is similar to the drinks to which they are related, so cyser should be matured as long as cider, and so on.

I have not spoken of distilling techniques in this volume, simply because it is illegal in this country to distill one's own spirits. If anyone did want to distill his own honey-based spirit then he would have to be aware that distillation has its pitfalls and what is produced can have toxic side-effects if not matured properly. That the ancient Celts distilled mead is certain, but their lore in this respect is lost to us. Distillation is best left to experts.

Recipes

All recipes here are for a gallon unless otherwise stated. Therefore, the demijon in all cases should be made up to a gallon of water.

With all recipes proceed as directed earlier in this booklet. As far as possible all recipes given here are traditional and use British honeys. As we are uncertain what yeasts and yeast nutrients were used by traditional brewers I have prescribed ones in common use.

1: Dry Mead

This is by far my favourite.

Three pounds of clover, heather or cowslip honey are preferred,, along with a quarter ounce of tartaric acid, one half ounce of malic acid and a fifteenth ounce of tannin. A steinberg yeast is said to produce good results. For dry mead the technique is to allow fermentation to continue until it ends naturally, thus ensuring that as much of the honey as possible is converted to alcohol.

2: Sweet Mead

This is very much a dessert drink.

Sweet meads can be made with exactly the same honeys as can dry meads, but the quantity of honey and the yeast will differ from those used in dry meads, as can the length of fermentation. Honey in quantities greater than three pounds will produce a surplus of sugar remaining after fermentation has finished. This will produce a strong, sweet brew. Alternatively, the fermentation can be stopped by the addition of campden tablets. This allows less honey to be used and produces a weaker, sweet drink. A third technique is to add honey to the must. Success has been achieved by adding one quarter pound of honey when the specific gravity reaches five. This process is repeated every time the five point is reached until the ferment has stopped.

3: Queen Elizabeth's Mead

A dessert drink said to have been favoured by Elizabeth the First.

Again three pounds of honey is used with a quarter ounce of tartaric acid, one half ounce of malic acid, nutrients, onre fifteenth ounce tannin and a madeira yeast. The ferment proceeds until three to sixth months when a bag of herbs is suspended in it. The bag contains one half ounce rosemary, one half ounce thyme and one quarter sweet briar. Tasting should occur daily until the herbal flavour is satisfactory and then the bag must be removed. Maturation for another six months and if necessary the use of finings to clear the liquor is required.

Meadowsweet can be used as a flavouring. This is a herb whose name derives from meadsweet. It was used by Saxon brewers to flavour their meads. It seems to have fallen from use, but a workable rule is that a pint of meadowsweet should be steeped in water and the resulting liquor included in the must.

4: Ale Mead

A neglected drink.

Take one pound of honey and boil it for forty five minutes with an ounce of hops and a little orange or lemon juice in six pints of water. Keep back a few of the hops until forty minutes, then add them. Strain the wort, add two pints of cold water and allow it to cool over night, then add the yeast. Treat this drink as beer, which means skimming it every day until time for bottling. The product may be drunk in a few months.

There is, however, much room for experiment here. Beer brewed flavoured with hops was a thirteenth century import to Britain. Prior to that time ale flavoured with a variety of bitter herbs was the staple drink. These herbs included rosemary, ground ivy, heather, bogbean to name but a few. Grout was a mixture of these. In more recent times spruce oil was used. This was said to produce much less of a hangover than hops do. An adventurous brewer has great room for experiment here. Interestinly, one ingredient of ale and, therefore, of ale mead would have been scurvygrass. This is a plant discovered by naturalists to contain substantial quantities of vitamin C. Before the introduction of citrus fruits it was brewed into an ale in place of hops etc, and was drunk every day to protect against skin ailments. An infusion of scurvygrass mixed with honey could produce a pleasant tonic containing vitamins and mineral salts.

5: Metheglynn

One recipe recorded from the nineteenth century but dating from earlier involves the following: Four pounds of virgin honey and two gallons of spring water, the white of an egg and the peelings of a lemon (orange will also be usable). These should be boiled for an hour and then rosemary, cloves, ginger and sweet briar added. After this is applied the yeast. After a few months the drink is said to resemble Tokay in its flavour. Interestingly, this was given as a braggon, a drink brewed with malt and bitter herbs. Yet no malt is found in the recipe, which suggests that the term braggon may have been variable in application and sometimes referred to a metheglynn flavoured with bitter herbs.

Metheglynn may be made by taking any mead recipe and adding herbal flavourings or spices. These were added seven days after fermentation had begun and left for four days, after which a racking took place and the fermentation was allowed to continue. Any combination of herbs may be added according to need or taste, but heather and elderflower produce a lovely taste. Cloves, cowslip, rosemary, nutmeg and ginger have also been used, along with many others whose healing powers are known to herbalists. In early times the quantities of herbs used were substantial, there being an ounce of whatever herbs were used for every gallon of brew.

The problem with metheglynn is the blending of flavours. Sometimes the herbs' flavour does not meld adequately with the flavour of the honey, creating a clash on the palate. For this reason it is advisable to be sparing with the herbs at first and to allow sufficient time to mature. There is no hard and fast rule here. Different herbs and combinations thereof will blend in varying degrees and over different lengths of time. Experience is the great teacher here.

Since the first edition of this bok came out I have experimented with dandelion metheglynn. Dandelion heads have a honeyish flavour. Unfortunately, if you use white sugar this honey flavour can hardly be tasted. I decided that the honey flavour would be augmented by using honey rather than white sugar. I took a pint of dandelion heads and soaked them overnight, straining them into the yet unyeasted must. The result is a potable metheglynn. I do not use it as table wine. Its use is partly medicinal, for dandelion is an anti-inflamatory herb which serves to alleviate joint pains, such as those of arthritis. Honey, being traditionally medicinal, enhances this effect. I recurrently suffer from a trapped nerve in the hip, the result of a minor rock-climbing injury many years ago. I have found that the pain disappeared after taking some of this metheglynn.

During the eigtheenth century a relation of metheglynn seems to have been used. It was customary to recommend that herbal infusions be taken mixed in ale and,therefore, it is likely that ale mead was as widely used as ale for this purpose. One recipe for back pain involved mixing a paste of comfreyroot with a posset of ale. A posset was a brew made from a an alcoholic drink, milk and spices or herbs, which were mulled or heated together and administered to the sick. Honey was often an ingredient of possets, in which it served to as a sweetener and an essential nutrient. Ale mead, a drink sweeter than ale, would have been a more pallatable brew to have with the bitter comfrey root. In general, herbs which could be taken in metheglynn could be taken in ale mead or ale. Fennel was extremely popular with the Saxons, who held it sacred and in possession of strong healing powers,so this too may be used. It is said to relieve wind and other stomach ailments.

6: Braggot

An ale-type drink.

Few recipes survive for braggot. The general rule is that the must requires a mixture of one pound of malt extract and one pound of honey which should be boiled together for fifteen minutes in seven pints of water. As with any malt drink the surface should be skimmed while the boiling is in process. Citric acid in the form of a little orange or lemon juice is necessary. When the wort cools ferment with an ale yeast. When fermentation is over rack for three months. It is said to be best served chilled in beer glasses. Some like it from pewter.

There is again room for experimentation here. It is not widely realized that ale was brewed not only from barley but from wheat, oats and rye. Modern barley-brewed and hop-flavoured beers are merely one limited element in a complex repertoire of drinks, many of which are neglected. It is possible that the malt for braggot came from any of these cereals, all of which can be used again by any home brewer capable of malting the grain. Similarly, any of the variety of bitter herbs mentioned above may be used to replace the hops. Experimentation here will produce many of the varieties of brew known to our ancestors.

7: Cyser

A neglected favourite.

This drink requires four pints of English apple juice and two pounds of honey. Mead acid mixture needs to be added. The yeast can be any yeast suitable for cider. An all purpose yeast will suffice. The honey is first dissolved in three pints of boiling water, then the apple juice is added. The must is made up to a gallon with cold water, twenty four hours later the yeast is added. It should be fermented until gravity drops to zero. Rack and store for three months and drink as cider.

The rules for the use of apples are those which apply to cider. A good cider is fermented from a blend of cider apples, cooking apples and crab apples, the precise mixture being determined by experience and taste. Dessert or eating apples do not make a good cider or cyser, although it has been suggested that James Grieves apples make a good cyser if used when freshly picked.

Cyser could be made in two ways. Cyder, a wine strength drink, used practically pure apple juice with a little water.Cider, known as ciderkin or small cider, used apple juice with much more water mixed in. You could mix honey into a cyder, although I feel that the more honey you are using the more water you will need to dilute it. Honey could be a flavouring or a minor ingredient of a cyder type drink, but it could be a major ingredient if the mead must of honey and water was mixed in more even proportions with apple. Similar drinks might be made using honey and pears, although I have never heard of their being made. What is of interest is that the monks used to use cyser as a vehicle for medicinal herbs, just as metheglynn was. It combined the goodness of apple with that of honey. There is room for experimentation with a wide range of herbal extracts here.

Traditional cider brewers used to hang a piece of steak or a dead rabbit over the brew, as this was felt to give it body. In Devon a dead rabbit was left in the vat. I know of no one who has tried this with cyser but, as with cider, it is probably traditional to drop a piece of meat into the brew. However, it is not necessary.

8: Melomel

A whole range of drinks here. Most white fruits can be used. I am concentrating on those using native British fruits.

A: Red mead. This uses four pounds of red currants and two pounds of honey, a polyflora, mixed flower, honey will suffice. Sometimes a half pint of white grape concentrate is added. Alternatively, an extra half pint of red currant juice may be used. The normal nutrients are required. Bordeaux yeast is said to produce good results. The red currants and honey are fermented together for three days and then the concentrate is added. Allow fermentation to continue to its natural end. A dry wine is produced.

B: Black mead. This drink uses four pounds of blackcurrants and two pounds of honey, which, as with Red Mead, can be a polyflora one. Along with these and the normal nutrients can be added one half pint of red grape concentrate, although before such concentrate was available more blackcurrants were probably used. One quarter ounce of malic acid is needed. Proceed as with Red Mead, but rack when the specific gravity reaches zero.

C: Gooseberry melomel. Gooseberries are a native British fruit and so were likely to have been used with honey. Although gooseberries can be used fresh, some brewers are of the opinion that the best melomel comes from those which have been allowed to over-ripen or go mouldy on the bush prior to picking. The sign that they are mouldy is that they split and allow mould to grow in the cracks. Whatever way is used, the best results are obtained from crushing the gooseberries and fermenting the pulp. I have obtained a delicious wine with a good body from gooseberries which I left in the must throughout fermentation, enabling their flavour to seep out. The recipe here uses six pounds of gooseberries and two pounds of honey. The honey is dissolved in six pints of water, along with the nutrients and a fifteenth ounce of tannin Then the crushed gooseberries are added. After twenty four hours the yeast, a sauterne or an all purpose yeast, is applied to the must.

Some brewers add a pint of white grape concentrate and a half pint of rose petals. Yellow or white petals seem popular. These add flavour but are an option They are allowed to steep in the must for three days, when racking must take place. At this point there are options. The must can ferment to dryness or, to make a sweeter product one quarter pound of honey is added whenever the specific gravity dips below five, until fermentation finishes.

D: Raspberry melomel. Raspberries were long thought to be a medicinal herb, whose juice was used as a palliative for colds and favours and was given to women in childbirth to ease the pain, so this brew would have been popular. Although they cannot be called a white fruit, they do not have the astringency of blackberries and elderberries. Raspberries are juicier than gooseberries, so only four pounds are needed. Any kind of honey will suffice, although good results have been obtained with lime blossom honey. Some melomel makers use a half pint of red grape concentrate, but extra raspberries will suffice. A red wine yeast will be needed, some swear by a port yeast and others think that madeira gives the best results. Two pounds of honey is dissolved in three quarters of a gallon of water and the crushed raspberries added. (Crushed fruit gives more body than mere juice). Campden tablets are inserted and the must is left for a day, after which the yeast is added. The whole is strained off after three days. At this point extra raspberry pulp or grape concentrate is added. If a dry melomel is required, leave the must to ferment to dryness. If a sweet one is wanted, add a quarter pound of honey every time the specific gravity reaches five.

9: Piment and Hippocras

Piments used a blend of grape juice and honey. The recipe is to simply mix white or red grape concentrate with two pounds of honey, along with the nutrients and a half ounce of malic acid. An all purpose yeast will suffice, but it is also possible to use yeasts suitable to the concentrate you are using. White piment requires a fifteenth ounce of tannin. The basic procedure is then followed.

Hippocras is a piment to which spices or herbs have been added. The rules for the addition of herbs and spices are exactly the same as for metheglynn. Cinnamon is, however, a popular spice which should be added at a quarter ounce per gallon. Cloves should be added but not kept in a hippocras, for they are best strained out after three or four days.

10: Honey Beer and Wine

A: Honey Wine. An old Kentish recipe involves boiling two pounds of honey in a gallon of water for an hour. During boiling upto half an ounce of ginger, cinnamon and cloves and a piece of bruised nutmeg are added. Alternatively the rind of a citrus fruit can be used. When the must has been allowed to cool the spices are strained away and an ale yeast is applied. Initially fermenting was in a wide vessel and scum was skimmed from the surface at intervals. After the initial ferment the must was transferred to a stone vessel where it was kept in a warm place until fermentation was over, when it was stored for upto six months in a cool place. It could then be drunk. This is really a light mead.

B: Honey Wine and Bitter Herbs. A drier recipe involved boiling a pound of honey in a gallon of water and adding bitter herbs, often but not necessarily hops. The procedure is identical to the one used above, except for the unusual practice of skimming off the bitter herbs and reboiling them for an hour. The resulting liquor is then used to top up the brew. This one needs to be kept for a year. Again it can be regarded as a light, cheap mead.

C: Honey Beer. This country drink must date from the nineteenth century when sugar became available in quantities. Boil two pints of water with a little ginger for thirty minutes. Then put either a pound of sugar and a half pound of honey (or one and a half pounds of honey if you want to use the older method) into a basin with the juice of three citrus fruits, lemons are preferred. Pour on four pints of cold water then add the boiling water. A teaspoonful of yeast should be spread on the surface. The basin should be covered in a cloth, ideally muslin for twenty four hours. After this it is strained through the muslin and bottled. It may be drunk a few days later. This is a light, cheap ale mead.

11: Caudle

This is a recipe popular in Elizabethan times, where it was used as a cold cure, a winter nightcap and was given in large doses to women in labour. It is also recommended for insomnia. Take a quarter pound of honey and a tablespoon of ground oatmeal. Pour over it two pints of brown ale. Use homebrew or real ale for the best quality and taste. It is then heated in a low oven for two hours. Keep the brew covered to avoid loss of alcohol. Do not let it get too near to boiling point, for alcohol evaporates at a temperature lower than water does. The juice is then strained off and nutmeg, some lemon juice and a glass of whisky or rum is added. The whisky or rum is probably post-Elizabethan, rum in particular became popular after the development of trade with the West Indies, which produced rum from molasses derived from the sugar cane.

Another form of caudle involved boiling strong ale, preferably homebrew or real ale, in one pan and a mixture of four level spoons of honey, some cloves and a mixture of spices in a pint of water in another. The two pans are then blended. Spirits and citrus fruit juice can be added according to taste.It should be noted that anyone willing to experiment with the wide variety of ale flavourings mentioned above can produce an equally wide variety of caudles.

12: Bishops

There is a vast range of bishops, in fact wealthy men often had their bishops brewed according to their taste. The general rule is that they are brewed of spices, citrus fruts and honey. Take a citrus fruit, an orange, lemon or lime. Make incisions into the rind and insert whatever spices you wish, although cinnamon, cloves, ginger and nutmeg are popular.Roast it in an oven. Put spices into a saucepan with a half pint of water and boil until half a pint remains. Pour in a bottle of heated red wine. Sometimes some of the

spirit is burned off with a lighted taper. Then add the roasted orange and leave it for ten minutes. Sometimes orange or lemon juice is then added. Strain the brew. Add the honey and extra spices if wished. It is then drinkable. If claret is used, it is called cardinal, and if champagne is used, the name is pope. An elderberry wine is good for bishops, for it has a strong taste that sets well against the honey.

The rule for bishops is that they are a brew of warmed wine with spices, citrus fruit and honey. Any bishop can be concocted using the techniques in the previous paragraph. Doctor Johnson liked one brewed from a little honey, a bottle of claret, four cloves and a half pint of water. To these he added a little brandy and curacao, and a pinch of nutmeg. Beware of pouring hot bishops into wine glasses. An old technique was to leave a spoon in the glass to take away the heat. Alternatively, drink from pewter.

13: Negus

This is a simple bishop.

A negus is made simply by heating a bottle of wine with a half pound of honey, a few cloves, up to six, and some lemon or orange juice. They are heated until near boiling in a quarter pint of water. Brandy or rum can be added, along with a little nutmeg. Elderberry wine is said to make a good negus. The elderberry is a much neglected fruit. It covers the hedges of England in Autumn, which is why the author of this book is often seen in early September cycling with large bags on his bicycle. If you ferment elderberry wine to dryness you get a wonderfully strong and fruity result. As elderberries cannot be used for melomels, they can certainly be used in bishops and neguses. Plum wine can also be used, as it is rich and strong.

It is possible to make a related drink by simply heating wine with honey and a few kitchen spices. There is no hard and fast rule for the honey, which should be added to suit one's taste. This is a kind of mulled wine. Traditionally wine was mulled by inserting a heated poker kept especially for the purpose. Mulled wines are a winter drink intended to be drunk hot.

14: Lamb's Wool

This is an ancient country drink. Core four large apples and fill the cores with honey. Sprinkle with nutmeg, although cinnamon always goes well with

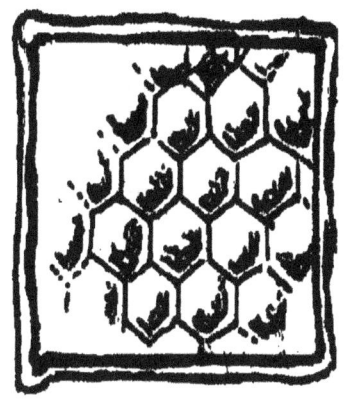

apple. Heat in a medium oven in a baking tin for three quarters of an hour. Pour four pints of strong homebrew over it. Heat gently, spooning the beer over the apples. Then you can strain off the liquid. The joy of this is that you eat the apples as well as drinking the brew.

A festal lamb's wool involved heating four gallons of ale with three pounds of honey and spices, sometimes nutmeg and ginger. The juice of four lemons was added. As a drink for winter parties this was ideal.

15: Wassail

This is a traditional drink of the midwinter festival and, later, Christmas. Its basis is an alcoholic drink, either ale or mead, to which is added whatever fruits and spices appeal to the individual's taste. The basic recipe is that to a mulled ale or mead should be added whole fruits and seasonings. In early times apples were probably added, along with possibly elderberries, blackberries,raspberries and bilberries, should any of these have been kept for the winter season. Bilberries would have been the most difficult to keep as they are picked in early summer. In more recent times raisins took the place of other fruits. Spices, such as cloves and cinnamon became available in more modern times.

One way of making wassail is to boil a half pound of honey and the spices (not fruits) in water for upto five minutes, add the juice of some citrus fruits and then pour in a bottle of wine or mead. Then heat in a slow oven until nearly boiling. Keep the lid of the vessel on to prevent evaporation of the alcohol. Serve hot.

There are no set wassail recipes. There is simply the rule that wassail should be hot, alcoholic, spicy, fruity and taken as a winter festal drink.

16: Athol Brose

This is a liquor very popular in Scotland, where it is thought to be the ideal breakfast drink. Essentially honey is mixed with a strong whiskey. There are no hard and fast rules for the mixture. Ideally the mixture is liquid enough to be drinkable. Too much honey would produce a drink too syrupy to be satisfying. Liquid honey should be used, as this dissolves more easily than solid honey. If the honey is solid, it should be lightly heated until it is melted.

17: Scotch Malmesey

This delicacy is in a class of its own. Take a gallon of ale wort (unfermented must) and add a gallon of water. To this take a pound of honey and a pound of sugar. Stir well. Make a yeast starter from a little of the must and some brewers' yeast. Ferment for a month. While this is in process soak a half pound of raisins, muscatel preferably, and two pounds of bitter almonds in a quart of whiskey. After the initial ferment is finished, strain off the almonds and raisins and add the whiskey to the must. You can bottle in a month's time and it is ready to drink at straight away. The raisins and almonds were kept for cooking, to which they brought some of the flavour of the whiskey.

18: Medicinal brews

Honey and vinegar. Various vinegars were were considered to have herbal qualities. Some were applied externally and so have no place in this book. Others were used in drinks, sometimes mixed with honey.

Cider apple vinegar and honey was long used to alleviate the symptoms of arthritis. As I pointed out earlier, cider can be brewed with the juice or the pulp, and the stronger brew is from the latter. However, only the former is now made commercially. I know this drink from personal experience, as my mother used it to alleviate arthritic pain and was convinced that there was some benefit. However, it is not a substitute for medical attention and seems to be a palliative for osteoarthritis, that caused by ageing, rather than the more severe forms such as rheumatoid arthritis.

Mix half a cup of vinegar with two tablespoons of honey and make up the cup with hot water.

Raspberry vinegar was used to cure colds and fevers and was very likely to have been administered with honey, as this would help soothe sore throats. As with the previous recipe mix half a cup with two tablespoons of honey and make up the rest with hot water.

Hedge Mustard was used as a cure for asthma. The sap of one of these plants was mixed with honey or treacle and given to sufferers. A mild drink could be made out of this by weakening the solution with water to make a sweet, syrupy concoction.

19: Mai Bowl.

This was originally a German drink based on the herb woodruff, a wild Maytime herb, but it found its way into Britain. Generally a handful of this herb was soaked in a pint of water, and liquor added to a mixture of white wine, which absorbed the flavour of the herb and champagne, which gave sparkle. In theory any sparkling wine will suffice. A light mead made with English honey, that is a pale honey, can replace the champagne. This practice involves the production of a mead with something of a sparkle about it, a process which involves allowing a secondary ferment to continue for a time after bottling. This is a delicate process which involves keeping the sparkling drink in strong bottles in a cool place, as any experienced wine maker knows how a ferment can blow the cork from a bottle if it is too strong. The English version of this drink used a good cyder instead of a white wine, and it is known as woodruff cup. Use equal quantities of mead and white wine with a handful of woodruff in a jug of water. However, do not soak the woodruff for too long, as the liquor will become too strong and this will destroy the delicacy of the taste. Avoiding this pitfall necessitates regularly tasting the woodruff until the required strength is reached. Then it should be mixed in with the mead and the wine.

20: Summer fruit drinks.

Fruits can be mixed in a great variety of ways with a great variety of wines or ciders. Honey had a role in the sweetening of such drinks. Some fruits, such as raspberries or gooseberries, can be somewhat tart, and it was customary to steep the fruit in a water sweetened with a little honey prior to mixing the resulting juice into a white wine or a cyder. The quantity of fruit should be sufficient to produce a strongly flavoured liquor. This would then produce a long drink, a kind of summer drink which has a relatively low alcohol content. This kind of drink is suitable for a time of year when people are apt to perspire, making them lose body fluids. Alcohol has a dehydrating effect, so highly alcoholic drinks are unsuitable for hot days. Low alcohol concoctions are, on the other hand, very apt indeed.

The Mystery of Heather Ale

That heather was used as a flavouring for alcoholic drinks is widely known. A drink of unidentified ingredients known as **heather ale** was known to the Picts. So highly valued was it that only selected individuals possessed the formula. When the Pictish kingdom fell to the Scots heather ale seems to have fallen from fashion, although it was known to the Danes who invaded Ireland in the ninth century. In fact three Danes, a father and two sons captured by the Irish after the battle of Clontarf, refused to trade the secret in return for their lives. They are believed to have been its last possessors. Yet in theory there is no secret. Heather flowers may be used to flavour ales and ale meads quite easily, being used just as hops or other bitter herbs were used. An infusion of heather leaves is quite easy to make, and it is just as easy to put the leaves into the must. The secret of heather ale is not, to my mind, how it was brewed, but what other ingredients went into it. The drink whose ingredients were so secret was probably a festal drink used for religious purposes and possessed of sacred significance. That the Picts had been Christian for many years before the secret died out has no importance. Many old customs survived long after the coming of Christianity, as people hedged their bets and honoured the old gods as well as the new.

When people die rather than reveal a secret it is likely to be important to them. Did heather ale contain special ingredients? One possibility is that it was a drink which had shamanic significance, a brew that enabled the tribal shamans to enter a trance state. Such shamanism persisted into the Christian era and went underground after the conversion of Europe to Christianity. Such brews would have involved the use of hallucinogenic mushrooms. The peoples of Northern Siberia preserved an ancient custom of brewing a mushroom wine. The basis of this drink was honey, which was boiled with mushrooms, which dissolve easily in boiling liquid. It is not impossible that heather, honey and hallucinogenic fungi were used to make a forgotten brew for ritual use. Certainly, the honey would have provided the sugar base without which the fungi, all of which are carbohydrate deficient, could not be fermented. The result was a brew which combined alcohol and hallucinogen in what was probably a heady concoction.

It is not impossible that honey was the basis for similar mushroom brews used across Eurasia in pagan times. At the midwinter festival it was the custom of shamans to enter the hallucinogenic state in which they felt themselves to fly the body into the other world. This was eagerly watched by the assembled tribe, who wished to know what the shaman could divine for them. A ferment of honey and hallucinogenic fungi would have been the stimulus for this quasi-religious experience. Perhaps this is the origin of the witches' cauldron known to us from folklore. It could well have been the pot in which such brews were distilled, along with other herbal concoctions.

However, a strong word of warning is appropriate here. These paragraphs are speculation, but what is not speculation is the dangerous ill-effects of addiction to such brews. In Siberia it is known that those addicted to the brew of honey and fly agaric quickly degenerate into incurable and shambling dementia. The muscarine, which produces the hallucinations, is a nerve poison. The fly agaric is bad enough, but it has utterly lethal relations, whose names - death cap, destroying angel and fool's mushroom - speak for themselves. These contain not only muscarine but also a blood poison. Nowadays a cure is possible, but it involves a long stay in hospital. This is not an area in which it is safe to experiment.

Rowan

Rowan was an old favourite with the Celts, though not the Saxons. It is a bitter berry completely unpalatable on its own, though rich in vitamin C. Even now it is used to make a jelly for eating with game. The ancient Welsh used to drop rowan berries in ale, in which it served as a bitter herb, and the result was a powerful brew. A drink used to be made from the vitamin C-rich rowan and used as a cure for skin ailments. This drink is bitter, so honey would have been needed as a sweetener. Furthermore, it was credited with magic powers to repel witches. In a prescientific age medicine involved elements of magic which have no place in modern practise. The ability to ward off witches would have been highly prized, and as rowan berries are nutritious it is likely that they were an ingredient of metheglynns. However, rowan is not a berry to be taken to excess, as I am not convinced that it is completely harmless if taken in great quantities.

Bibliography

Country Wines, Mary Aylett, Odhams Press 1953
Making Mead, Brian Acton and Peter Duncan, Amateur Winemaker, 1970
500 Recipes for Homemade Wines and Drinks, Marguerite Patten, Hamlyn 1978
Sunday Times, 20 Dec 1992, for information on shamanism at the midwinter festival.

Index

Dry Mead	14
Sweet Mead	14
Queen Elizabeth's Mead	15
Ale Mead	15
Metheglynn	16
Braggot	17
Cyser	18
Melomel	18
Piment and Hippocras	20
Honey Beer & Wine	20
Caudle	21
Bishops	21
Negus	22
Lamb's Wool	22
Wassail	23
Athol Brose	23
Scotch Malmesey	24
Medicinal brews	24
Mai Bowl	25
Summer fruit drinks	25
The Mystery of Heather Ale	26
Rowan	27